First Responder Police Journal Notebook
To Serve And Protect First Responder Journal Series Gift Book For Police
Paperback ISBN: 978-1-989733-41-7
Copyright Dunhill Clare Publishing 2020
All Rights Reserved. Cover Design by Sharon Purtill

www.ingramcontent.com/pod-product-compliance
Lightning Source LLC
Chambersburg PA
CBHW071719020426
42333CB00017B/2332